IMAGES
of America

THE NAVY IN PUGET SOUND

On the Cover: In the years before World War I, the USS *St. Louis* (C-20) was often in limbo. The U.S. Navy retired the protected cruiser from active service in the Pacific fleet in 1910, but the ship remained in reserve at San Francisco and in the waters of Puget Sound. In this image, the *St. Louis* undergoes maintenance at Puget Sound Navy Yard near Bremerton, Washington, c. 1913. (Puget Sound Navy Museum.)

IMAGES
of America

THE NAVY IN PUGET SOUND

Cory Graff
Puget Sound Navy Museum

ISBN 978-0-7385-8081-4

Published by Arcadia Publishing
Charleston SC, Chicago IL, Portsmouth NH, San Francisco CA

Printed in the United States of America

Library of Congress Control Number: 2010921322

For all general information contact Arcadia Publishing at:
Telephone 843-853-2070
Fax 843-853-0044
E-mail sales@arcadiapublishing.com
For customer service and orders:
Toll-Free 1-888-313-2665

Visit us on the Internet at www.arcadiapublishing.com

Contents

Acknowledgments

This book would have not been possible without the Puget Sound Navy Museum. Special thanks goes to the institution's volunteers and staff, particularly Lindy Dosher, Danelle Feddes, Heather Mygatt, and Kathrine Young. Images for this volume came from a number of sources, including the United States National Archives in Washington, D.C., the Naval History and Heritage Command, Grumman History Center, Eleanor Wimett at Patriot's Point Naval and Maritime Museum, Greg Hagge at Fort Lewis Military Museum, Kathleen Crosman at the Pacific Alaska Region of the National Archives, and Carolyn Marr at Museum of History and Industry. Thanks to individuals Glenn Humann, Jim Caldwell, and Steve Pickens, who loaned photographs for this volume. Others helped with advice, editing, and their artistic skills, including Calvin Graff, Gary Graff, P. J. Muller, Mike Lombardi at the Boeing Company Archives, Katherine Williams, Leo Echaniz, Adrian Hunt at the Flying Heritage Collection, Tracy White, Steve Ellis, Dennis Parks, Tim Detweiler, and Shauna Simon. And special thanks to Sarah Higginbotham, Michael Stafford, Ryan Easterling, and Devon Weston of Arcadia Publishing, who helped make this book a reality.

INTRODUCTION

On May 2, 1841, U.S. Navy vessels USS *Vincennes* and *Porpoise* sailed into the Strait of Juan de Fuca at the mouth of Puget Sound. The men of the United States Exploring Expedition had been traveling the globe for nearly three years, and most of them were ready to sack their commander, the overbearing Charles Wilkes. The new task ahead was a pleasant one, however, compared to what they had been through. The exhausted crew had battled massive walls of floating ice in the frozen reaches of Antarctica, climbed to the top of an active volcano in Hawaii, and fought with the natives of Fiji. Now all they had to do was survey paradise.

At least it was paradise as far as the navy was concerned. Like Britain's George Vancouver before him, Wilkes and his contemporaries saw the Puget Sound as an ideal western port for the warships of America's growing navy. Miles of deep-water inlets and sheltered shores, along with a nearly inexhaustible supply of natural resources, made the region extremely valuable. And unlike some of the groups the explorers had encountered in the Pacific, the natives were friendly.

As more white settlers came to the area in the years after Wilkes, the mood of the area's original inhabitants changed. By 1855, the sloop of war USS *Decatur* was dispatched from Hawaii to protect the men and women of the tiny village named Seattle from the increasingly restless Native Americans. Tensions boiled over in January 1856, when hostile forces attacked the outpost on the shores of Elliott Bay. The ship's marines and civilian volunteers fought back, assisted by repeated and terrifying fusillades from the *Decatur's* heavy guns. The very presence of the navy warship and its crew saved the citizens of Seattle from a terrible massacre.

When it was clear that Puget Sound would surely be part of the growing United States of America, navy men traveled west, looking to establish permanent bases in what was dubbed "America's Mediterranean." Lt. Ambrose Barkley Wyckoff came to Washington Territory to further explore and document the sound, with an eye toward building a naval base and shipyard. Back east, in the other Washington, Wyckoff became a strong advocate for the navy's presence in Puget Sound.

In 1891, carrying $10,000 awarded by Congress, Wyckoff returned to the newly formed state of Washington to purchase land for a new navy yard on the shores of Puget Sound. He and others on the shipyard committee chose a site at Point Turner, on Sinclair Inlet, near what is today the city of Bremerton. In a short time, the navy held slightly more than 190 acres for the purchase price of $9,587.25. Wyckoff became the first commander of Puget Sound Naval Station. By 1903, the shipyard facilities and construction efforts had made the base the largest employer in the region.

While Puget Sound offered an exemplary safe harbor for the repair and maintenance of navy ships working in the vast Pacific Ocean, it would also become the birthplace of new vessels. The Puget Sound Navy Yard's (PSNY) first navy craft could only be called modest. It was a small water barge, slid sideways into a flooded dry dock basin in 1904. That same year, the Moran Brothers Company, near downtown Seattle, launched the 441-foot-long hull of a Virginia-class battleship.

Though it was to be named USS *Nebraska* (BB-14), citizens of the region thought of the new vessel as Seattle's battleship. They had donated over $100,000 to the project to assure that the ship was built in Puget Sound. In 1908, the *Nebraska* joined the "Great White Fleet" in San Francisco, traveling more than 29,000 miles around the globe.

America's Great White Fleet visited Puget Sound too. On the first leg of their journey, the 16 battleships and their support vessels sailed from Virginia to San Francisco by steaming around Cape Horn at the bottom of South America. The third part of the voyage would be from San Francisco to the Philippines. In between, the ships cruised north along the West Coast to Washington State, solely to visit the growing Puget Sound region.

It was a proud moment for Washingtonians. In May 1908, the beautiful white-hulled battleships arrived. Amid the parades, speeches, and celebratory dinners, each crew from every battleship was given a black bear cub mascot from the city of Aberdeen, Washington, as a gesture of goodwill and good luck on their cruise around the world.

In the years that followed, the Puget Sound became a regular port of call for the U.S. Navy. Many of America's famous warships would visit the region during their duties in the Pacific, including veterans of the Spanish-American War such as the USS *Oregon* (BB-3), the famous battleship USS *Arizona* (BB-39), and the navy's first aircraft carrier, USS *Langley* (CV-1). In 1928, the nation's two newest aircraft carriers, USS *Lexington* (CV-2) and USS *Saratoga* (CV-3), were both at Puget Sound Navy Yard at the same time, docked side by side.

The *Lexington* returned a little more than a year later to help one Puget Sound city out of a crisis. A rare drought in the winter of 1929 left Tacoma without enough water in local reservoirs to generate hydroelectric power. At Baker Dock, resourceful engineers plugged the "Lady Lex" into the electrical grid, using the carrier's boilers to help Tacoma through a tough Christmas.

The oldest navy warship to ever visit the Puget Sound was built in Boston and launched in 1797. USS *Constitution*, still a commissioned warship today, visited the region during a three-year tour of American ports after its restoration in 1930.

Aviation, too, became a critical part of the navy's presence in Puget Sound. Famous flyers Eugene Ely and Hugh Robinson dazzled Seattle residents as they buzzed over Elliott Bay in their Curtiss airplanes in 1911. Years later, William Boeing built his first aircraft in hopes that he could sell the design to the navy as a training tool for new floatplane pilots. As the nation ramped up for World War I, Boeing's first contract for building aircraft was awarded by the navy. Later the Boeing Airplane Company built a number of premier fighter aircraft for the navy that would fly from carriers in the 1920s and 1930s.

Facilities for accommodating aircraft soon followed, creating an aerial umbrella for the ports and yards in the Puget Sound region. In the early 1920s, the navy and army jointly established primitive flying facilities at Sand Point, on the shores of Lake Washington. On November 22, 1928, the field was officially named Naval Air Station Seattle.

In the shipyards, the navy was using Puget Sound shipbuilders to add to its fleet. Both the Moran Company and Puget Sound Navy Yard assembled a fleet of submarines in the 1910s. Designed elsewhere, these underwater craft were often called "pigboats" by their wary crews. A pair of these Seattle-built subs was secretly smuggled northward to Canada on the eve of World War I, violating America's neutrality laws.

At Puget Sound Navy Yard, the shipfitters built pioneering support vessels for the growing navy. USS *Pyro* (AE-1), named after the Greek word for "fire," was America's first navy ammunition ship and was launched in 1919. The USS *Medusa* (AR-1), the navy's first purpose-built repair ship, slid down the ways at PSNY four years later. Both vessels were moored in Pearl Harbor during the Japanese attack in 1941.

New warships, too, sailed out of Puget Sound on their shakedown cruises. In Tacoma, the USS *Omaha* (CL-4), lead ship in its class of light cruisers, slid down the ways at Todd Dry Dock and Shipbuilding Company in 1920. A decade later, the heavy cruiser USS *Louisville* (CA-28) emerged from PSNY. Both ships went into combat in World War II, the *Omaha* assisting in the capture of the German raider *Odenwald* in the Atlantic and the *Louisville* fighting in the Pacific for nearly the entire conflict.

World War II came early to the U.S. shipyards and factories. As nations in Asia and Europe clashed, America worked to assist its allies. Sorely needed improvements and expansions followed at naval bases and facilities in the Puget Sound area, not only to help Britain and France, but also to bolster the United States for the war that seemed to be on the horizon.

After Japan's attack on Pearl Harbor, western Washington was viewed as a potential target for future Japanese aggression. Working frantically, the region assembled a complex defensive web of submarine nets, barrage balloons, smoke generators, antiaircraft guns, and bomb shelters in order to protect navy, army, and military production assets in the area. Valuable targets were simultaneously hidden and reinforced for the expected attack that never materialized.

In general, Puget Sound's various installations took on specific jobs during the war effort. Seattle-Tacoma Shipbuilding Corporation (STSC), a joint venture of Todd and Kaiser Shipbuilding, became expert at creating escort carriers in its bustling Tacoma shipyard. The small ships could be built fast and at a fraction of the price of large fleet vessels. Tacoma's "baby flattops" served as convoy escorts in the Atlantic and fought in many pivotal battles in the Pacific.

On Harbor Island near Seattle, STSC's construction facilities turned out a long line of sleek destroyers with a workforce of over 17,000 men and women. The Seattle Division delivered 126 new ships in one 36-month period from 10 busy slipways that were running 24 hours a day.

The navy yard at Bremerton also churned out destroyers and destroyer escorts while taking on the massive job of servicing, upgrading, and repairing the multitude of warships fighting in the Pacific. Damaged battleships raised from the muddy bottom of Pearl Harbor were transformed—outfitted and modernized—at PSNY. Over time, these "Pearl Harbor Ghosts" were sent back into the fight by Washington's dedicated army of shipwrights working day and night. When kamikaze suicide aircraft struck the fleet in the last years of the war, it was often Puget Sound Navy Yard that would receive the twisted steel decks of the damaged carriers for quick repairs.

While the carriers were laid up in Bremerton, their fighter and bomber aircraft would fly from Naval Air Station Seattle and outlying airfields. From Puget Sound runways, naval aviators could continue to train and prepare to once again go back into combat.

One of the navy's only losses within Puget Sound came in August 1943 when the USS *Crow* (AMc-20) was accidentally sunk by U.S. Navy aircraft. Towing a target tug for practice torpedo runs, the wooden-hulled *Crow* was struck by an errant weapon. The torpedo punched in the side of the converted trawler below the waterline and sent it to the bottom. Soaked, but unhurt, the entire crew of the small vessel survived the mishap.

Smaller facilities around the Puget Sound region also contributed to the war effort. Numerous mid-sized shipyards in Everett and on Lake Union and Lake Washington turned out support vessels for the U.S. Navy and U.S. Coast Guard. Radio facilities on Bainbridge Island helped intercept and decipher Japanese codes. Keyport became the home of the navy's torpedo school, and the remote town of Bangor was established as a naval weapons magazine. In the world of aviation, Whidbey Island grew into a major air base for land and sea-based patrol bombers while both the towns of Arlington and Shelton hosted important auxiliary airfields for navy warplanes.

The Coast Guard, under the command of the navy during wartime, shared the daunting task of protecting it all, from the wave-tossed miles of open ocean off the Strait of Juan de Fuca to the piers crowding Seattle's dimly lit waterfront. Both men and a number of women joined the Coast Guard ranks during World War II.

Thousands of soldiers and sailors passed through Puget Sound on their way to combat in the Pacific. Scores of civilians came to the region too, looking to fight the war in the shipyards and factories. Men and women who had never before seen Washington's snow-covered mountains and picturesque waterways fell in love with the Pacific Northwest. During the war and after the fighting ended, they settled in cities and growing suburbs, making homes in the place they so greatly admired.

The end of World War II may have slowed the navy's growth in the region temporarily, but there was still work to be done. Some older vessels were towed home and torn apart for scrap. Many others were simply put in reserve. The shipyard at Bremerton famously became the resting spot for retired vessels, including battle-tested Essex-class aircraft carriers.

Years later, when fighting broke out in Korea, some of these ships were brought back into service to assist United Nations forces. But the "mothballed" aircraft carriers needed much more than a fresh coat of paint. Naval aviation technology had advanced greatly, and new jet-powered aircraft were faster and heavier. "Modernization" meant massive changes taken on by legions of workers, including those at Puget Sound Naval Shipyard.

Carriers like the USS *Essex* (CV-9) and *Lexington* (CV-16) departed Washington waters for combat near Korea or service with the fleet, equipped with revamped catapults, improved elevators, and jet-blast deflectors. Most notable were the ships' new, stronger, angled flight decks, which allowed jet planes to safely fly back into the air after a missed approach.

Well over 150 years after the first ships reconnoitered the forest-lined waterways of the region, the navy remains an important part of the Puget Sound community. In both peace and war, the U.S. Navy and Washington's citizens have shared an enduring partnership that lasts to this day.

One

Beautiful Waters

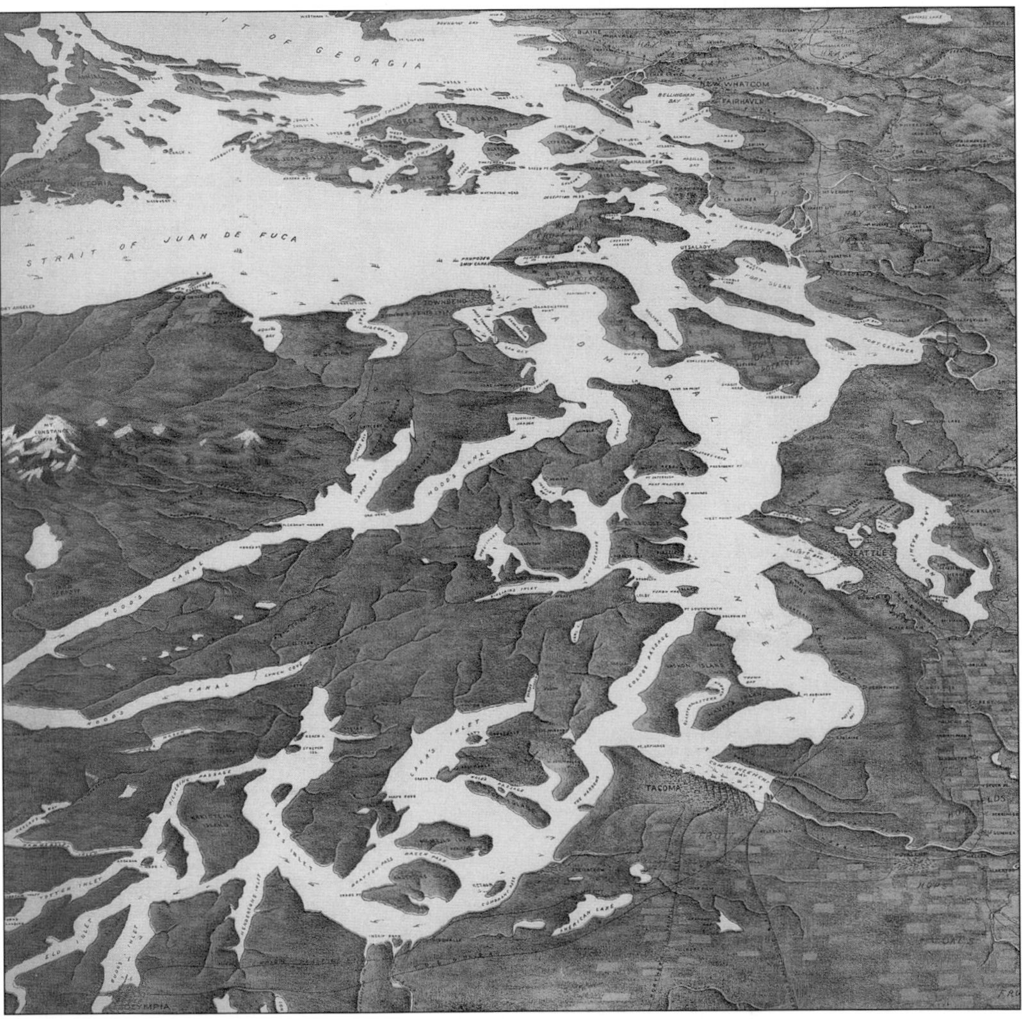

This 1891 bird's-eye-view map, printed by a Seattle lithography company, shows nearly every twist and turn in the complex topography of Puget Sound. With the sound's multitude of wondrously deep-water inlets and bays, it was only a matter of time before the U.S. Navy established a strong and enduring presence in what this mapmaker dubbed "The American Mediterranean." (Author's collection.)

The first American exploration of Puget Sound was led by U.S. Naval officer Charles Wilkes in 1841. The United States Exploring Expedition had traversed great expanses of the Pacific and the seas surrounding Antarctica before investigating the West Coast of the United States nearly three years into the voyage. Wilkes wrote of Puget Sound, "Nothing can exceed the beauty of these waters and their safety. Not a shoal exists . . . that can in any way interrupt their navigation by a 74-gun ship. I venture nothing in saying there is no country in the world that possesses waters equal to these." (Author's collection.)

The first navy ship to sail around the world, USS *Vincennes* went on to serve as Wilkes's flagship during the United States Exploring Expedition. In the summer of 1841, the 18-gun sloop of war entered Puget Sound and assisted in extensive surveys—the first conducted by the U.S. Navy. The explorations took place nearly a half-century after a British Royal Navy expedition, led by George Vancouver. (Naval History and Heritage Command.)

On January 26, 1856, the U.S. Navy sloop of war *Decatur* fired at a band of attacking Native Americans led by Chief Leschi. The exploding cannon balls, which killed an unknown (but small) number of the natives, kept the tiny settlement of Seattle from being overrun and its white residents massacred in what would become known as the "Battle of Seattle." This image of the *Decatur* or its sister ship was taken years later. (Museum of History and Industry.)

As USS *Decatur* prepared to defend Seattle, a naval officer, T. S. Phelps, made this sketch of the fledgling village from the deck of the ship. A far cry from the large city it would one day become, the tiny enclave was surrounded by trees and centered around Henry Yesler's sawmill. (Museum of History and Industry.)

In 1877, navy lieutenant Ambrose Barkley Wyckoff was tasked with surveying Puget Sound. Soon afterwards, he began to lobby in Washington, D.C. for the establishment of a shipyard in the sheltered waters of what would become Washington State. In 1891, Wyckoff returned with money from Congress to select a spot for the navy yard. He paid $9,587.25 for slightly over 190 acres near Point Turner on Sinclair Inlet. (Puget Sound Navy Museum.)

The storm-battered USS *Nipsic* carried out its final days at Puget Sound Naval Station. Sailors built a roof over the sailing ship's weakened hull and used the vessel as a receiving ship and floating prison. Whether *Nipsic* held new recruits or salty troublemakers, sentries posted at the gangways could easily discourage escape. The ship was sold in 1913 and converted into a barge by its civilian owners. (Author's collection.)

The navy viewed Puget Sound as an ideal place for its big ships. Miles of sheltered waterways with deep-water access allowed for the best ports north of the Bay Area. This image, taken in June 1899, shows the battleship USS *Iowa* (BB-4) in the first dry dock built near the growing city of Bremerton. The long narrow basin could be pumped free of seawater, creating a platform for maintenance, repair, and even construction of large oceangoing vessels. (Puget Sound Navy Museum.)

Puget Sound Navy Yard (PSNY) later became the birthplace of submarines, destroyer escorts, and even a heavy cruiser. But the first vessel built in the yard was much more modest. A navy water barge, launched in 1904, was the first major project completed at the navy base. Here the barge is side launched into the full dry dock basin on May 26, 1904. (National Archives/U.S. Navy.)

USS *Philadelphia* (C-4) had been a flagship in both the Atlantic and Pacific, carrying admirals and steaming across the world's oceans. The protected cruiser was ordered to Bremerton in 1902, just 12 short years after it was commissioned. *Philadelphia* went the way of the *Nipsic*—becoming Puget Sound Navy Yard's training and prison ship. The vessel's crew wrangled an unusual mascot, a goat named Teddy, who often wandered the *Philadelphia's* covered decks. (Author's collection.)

This protected cruiser was the first major navy ship to be named after a Puget Sound city. Launched in 1892, USS *Olympia* (C-6) was built in San Francisco but carried the namesake of Washington State's capital city into combat during the Spanish-American War. The vessel is today a museum ship moored in Philadelphia, Pennsylvania. (Author's collection.)

This tattered postcard shows a heavily retouched image of USS *Tacoma* (C-18) anchored near the city of Tacoma on Commencement Bay. After commissioning in California in 1904, the protected cruiser steamed up the West Coast and briefly visited the city of its namesake before setting off for Hawaii. After this brief stint in the Pacific, the *Tacoma* served for years in the Atlantic and Caribbean until it was lost near Mexico. (Author's collection.)

USS *Nebraska* (BB-14) was the first and last battleship to be built on the West Coast north of San Francisco. The citizens of Seattle donated thousands of dollars to the project in order to convince the navy to award the contract to the Moran Brothers Company. The project was begun in 1902. The photograph above was taken during the *Nebraska's* fitting out process after 1904. The photograph at left was taken in June 1906, more than a year before the battleship's commissioning in the navy. (Both, National Archives/ U.S. Navy.)

This famous image of the Moran Company shipyard captures the passenger steamer *Orizaba* in the foreground with USS *Nebraska* (BB-14) moored behind. The keel of the Virginia-class battleship was laid down in 1902 under the roof of the open building seen on the left, called "Ship Shed A." This *c.* 1907 image shows the vessel nearly complete, with masts, superstructure, and triple funnels assembled and installed. (Author's collection.)

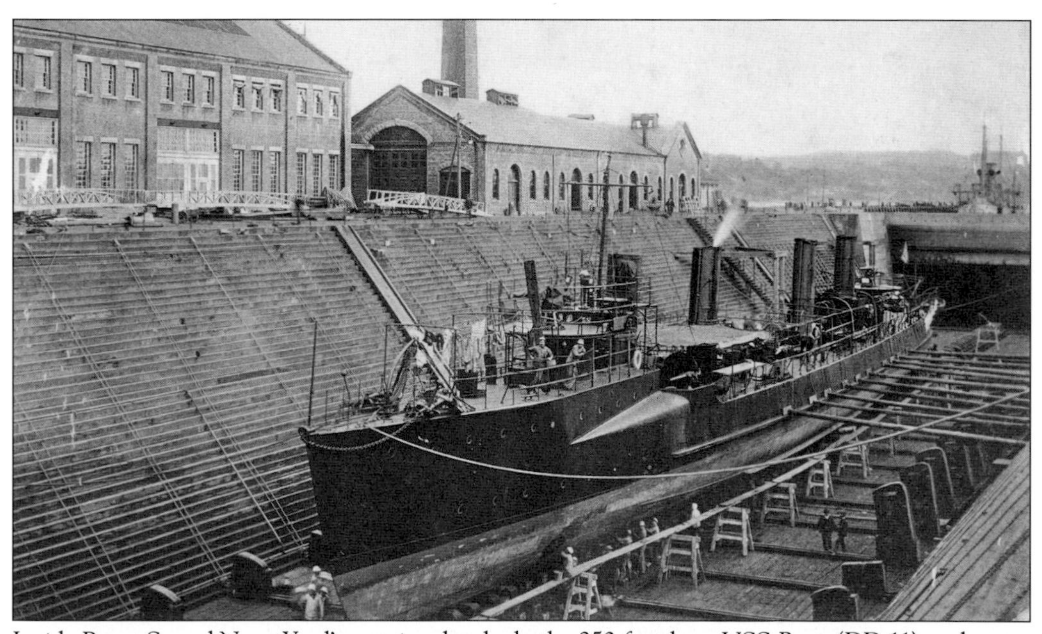

Inside Puget Sound Navy Yard's massive dry dock, the 250-foot-long USS *Perry* (DD-11) undergoes maintenance around 1906. That same year, the crew of the destroyer worked to keep the peace in the city of San Francisco after the devastating earthquake. *Perry* was a Bainbridge-class vessel—one of America's first destroyers. Some "old salts" considered the appearance of the ships downright hideous, with multifaceted hulls, low-slung decks, and oddly spaced funnels. (Author's collection.)

On a sunny day on Puget Sound, sailors from the USS *South Dakota* (ACR-9) paint the hull of their armored cruiser. The brilliant white hulls of U.S. Navy ships of the era needed to be continually maintained. This image was taken near Bremerton in May 1908. (Puget Sound Navy Museum.)

Two

The White Fleet

Prior to the arrival of the Great White Fleet in Puget Sound, an impressive collection of armored cruisers led the battleships as pathfinders. This composite image shows some of the ships in the spring of 1908 off Bremerton. From left to right, they are the USS *Tennessee* (ACR-10), *Washington* (ACR-11), *Pennsylvania* (ACR-4), *Wisconsin* (BB-9), *California* (ACR-6), and *West Virginia* (ACR-5). The battleship *Wisconsin* was at PSNY for recommissioning. (Puget Sound Navy Museum.)

On May 21, 1908, the U.S. Navy's battle fleet arrived at the Strait of Juan de Fuca. Enthusiastically welcomed by the citizens of Puget Sound, the Great White Fleet of 16 battleships and various support vessels took a few days off from its circumnavigation of the globe to be honored with parades, dances, and barbeques. When Admiral Sperry received the key to the city of Seattle, he told a reporter: "I have found Puget Sound a delightful place. If I were a millionaire I believe that I would give half of my possessions for the experience of going through the Sound. It is a beautiful body of water, and I have enjoyed every moment of my visit here." (Author's collection.)

Famous Pacific Northwest photographer Asahel Curtis leaned over the side of the Seattle Hotel to shoot this image of the parade in celebration of the Great White Fleet on May 26, 1908. About 300,000 people turned out to watch soldiers, sailors, militia, and fraternal orders march up Second Avenue. The sailor who sent this postcard home to his parents in Nebraska claimed that if they looked closely, they could see the 16 bear cubs given as mascots to each battleship in the fleet by the city of Aberdeen, Washington. (Author's collection.)

A steam-powered motor launch from the USS *Milwaukee* (C-21) was photographed near the Puget Sound Navy Yard in the summer of 1908. The protected cruiser *Milwaukee* had been put in the reserve fleet at Bremerton that year. The ship was later lost in 1917 during attempts to free the stranded submarine *H-3* near Eureka, California. (Puget Sound Navy Museum.)

The sleek-looking USS *Goldsborough* (TB-20) was photographed at Puget Sound Navy Yard in 1908. Laid down in Portland, Oregon, at Wolfe and Zwicker Iron Works in 1898, the torpedo boat was finished and commissioned in Bremerton on April 9, 1908. Though it sailed from California and Oregon ports for much of its navy career, the *Goldsborough* returned to Puget Sound for decommissioning in 1919. Months later, the vessel was scrapped. (Puget Sound Navy Museum.)

Excavating tons of dirt via railcar, crews work to create what will become dry dock two at Puget Sound Navy Yard on May 1, 1910. The building process took more than three years, from 1909 to 1912. In the background, the USS *Tennessee* (ACR-10) can be seen in the yard's smaller dry dock one. (Puget Sound Navy Museum.)

On the stepped sides of a dry dock at Puget Sound Navy Yard, shipfitters pose, many holding their rivet hammers, in this undated photograph. Wrangling heavy steel plates and hammering rivets was dangerous and strenuous work taken on by men and women of all ages. One can see a number of school-aged boys among the group. (Puget Sound Navy Museum.)

USS *Pennsylvania* (ACR-4) and USS *Charleston* (C-22) are moored side by side at Puget Sound Navy Yard in 1911. A postcard version of the image includes a comment from a shipyard worker: "The crane is 100-ton capacity and when visitors ask what it is, the sentry tells them it's an airship landing (mast). I put in two months on that crane driving rivets." (Puget Sound Navy Museum.)

The armored cruiser USS *Pennsylvania* (ACR-4) receives minor repairs and a new coat of paint at Puget Sound Navy Yard around 1911. The *Pennsylvania* is perhaps most famous as the landing platform for barnstorming pilot Eugene Ely in January of that year. Later in 1911, the vessel was transferred to the reserve fleet based in Puget Sound. In 1912, the cruiser was renamed USS *Pittsburgh* in order to free the name "Pennsylvania" for a newly planned battleship. (Puget Sound Navy Museum.)

A postcard scene shows some of the activities at Puget Sound Navy Yard around 1911—from tiny A-class submarines to temporarily stored cruisers and their motor launches. Many of the retired vessels, technically put in reserve, were sent back into active service as the nation rearmed for World War I. (Author's collection.)

Flitting around USS *West Virginia* (ACR-5), daredevil pilots Eugene Ely (above) and Hugh Robinson dazzle sailors and civilians alike with their noisy Curtiss flying machines. The pair is famous in the lore of naval aviation. Ely made the first takeoff from a ship in 1910, from a platform built over the deck of USS *Birmingham* (CL-2). In January 1911, using Robinson's makeshift arresting gear, Ely made the first landing on the deck of USS *Pennsylvania* (ACR-4). That July, the flyers of the Curtiss Exhibition Team buzzed over Elliott Bay, with Robinson making splashy landings and takeoffs with his new "hydroplane" aircraft. (National Archives/U.S. Navy.)

Never sticklers for accuracy, postcard publishers claimed this was an image of the submarine USS *Pike* (SS-6), diving in Puget Sound. Never mind that other, nearly identical versions of the card claim the photograph was taken near San Diego. Regardless, the *Pike* spent more than two years based in Puget Sound in reserve before being transferred to the Philippines in early 1915. (Author's collection.)

After its construction at the Moran Company, submarine *F-3* is put through its paces near Port Townsend in the summer of 1912. The 330-ton vessel was the first of nine submarines built in Seattle, seven of which served in the U.S. Navy. After a collision with another submarine in 1917, the vessel was used for training and in underwater photography experiments before being decommissioned in 1922. (Author's collection.)

USS *Oregon* (BB-3) was a veteran of the Spanish-American War. These images, taken in 1912, show the "old battlewagon" at Puget Sound Navy Yard after it was recommissioned in 1911. The battleship became a floating museum moored in Portland, Oregon, in 1925, but was converted into an ammunition barge during World War II. The remains of the *Oregon* stayed moored at Guam for many years. In the 1950s, the battered carcass of the vessel was finally scrapped in Japan. The mast and funnels of the ship are preserved at parks in Portland, Oregon. (Both, Puget Sound Navy Museum.)

When facilities in British Columbia were not available for yearly maintenance, the British cable ship *Restorer* came to Puget Sound Navy Yard. Seen here, the water rushes into the dry dock on February 14, 1912, as the ship prepares to return to the job of stringing underwater telephone lines off the coast of western Canada. When war came, *Restorer* continued its work with an American registry. (Puget Sound Navy Museum.)

The services of the Puget Sound Navy Yard were not reserved exclusively for U.S. Navy ships. Here the U.S. Army transport *Dix* is overhauled in dry dock two on May 1, 1913. The *Dix* was a frequent visitor to Bremerton. A photograph of the hold of the cargo ship, shot in 1915, reveals a store of 14-inch guns, most likely part of the ongoing job to re-gun battleships at the shipyard. (Above, National Archives/U.S. Navy; below, Puget Sound Navy Museum.)

A pair of warships shares dry dock two during recommissioning at Puget Sound Navy Yard in the summer of 1913. The protected cruiser USS *Milwaukee* (C-21), in the background, would become part of the Pacific Reserve Fleet. The monitor USS *Cheyenne* (M-10), formerly USS *Wyoming*, was being converted to a submarine tender. (National Archives/U.S. Navy.)

The *Iquique* and its sister submarine, *Antofagasta*, have an unusual history. Launched in Seattle in 1913, the pair was destined for the Chilean navy until the deal fell through. On the eve of World War I, the craft were "snuck" northward in Puget Sound under darkness, acquired covertly by the Province of British Columbia, and then transferred to the Royal Canadian Navy. Renamed CC-*1* and CC-*2*, the vessels became Canada's first submarines. (Jim Caldwell.)

This rare image shows both Seattle-built "F-boats"—*F-3* and *F-4*—moored in Honolulu, Hawaii. Built by the Moran Company, the ships were commissioned in 1912 and 1913 respectively. *F-3* spent much of its career serving along the West Coast while *F-4* served in Hawaii. They were reunited briefly in late 1914 when *F-3* participated in exercises in Hawaiian waters. On March 25, 1915, *F-4* sank during a test dive off Honolulu. All 21 crewmen on board the submarine perished in the accident. (Author's collection.)

H-3 (originally named the *Garfish*) was built by the Moran Company in Seattle from 1911 to 1914. This image was taken in Puget Sound shortly after the submarine was commissioned in March 1914. In late 1916, the vessel ran aground in Northern California. After four months of troublesome salvage efforts, the submarine was transported to a nearby bay on huge wooden rollers and relaunched. (National Archives/U.S. Navy.)

An old veteran, near the end of its career, shares the dry dock with USS *Oregon* (BB-3) in 1914. USS *Concord* (PG-3) was launched in 1890 and fought in the Battle of Manila Bay in 1898. The former gunboat would soon be moored at Astoria, Oregon, as a quarantine barge for the Public Health Service. (Puget Sound Navy Museum.)

A pleasant cruise near Puget Sound Navy Yard does not sound too bad—but this image was taken in February. On the icy waters of Sinclair Inlet, unlucky detentioners get "boat drill" duties in the winter of 1915. Detentioners were new recruits who were quartered separately for a few weeks after arrival in order to prevent the spread of disease. In the background lie Puget Sound Navy Yard's dry docks, support buildings, and the masts and superstructures of various visiting warships. (Puget Sound Navy Museum.)

Side by side in dry dock, two Puget Sound Navy Yard–built submarines can be compared. *H-6*, a type designed before World War I, can be seen on the left. *O-2*, seen here only four days after it was commissioned, was a bigger vessel that could stay at sea on longer patrols. This image was taken on October 23, 1918. (Puget Sound Navy Museum.)

The commandant of the Puget Sound Navy Yard, Robert E. Coontz (center), poses for a photograph in front of the USS *Wyoming* (BB-32) in 1918. Orderlies Jerry Blake (left) and marine sergeant John Delaney stand at Coontz's side. (Puget Sound Navy Museum.)

On a typical rainy spring day in 1918, submarine *O-2* stands ready for launch at Puget Sound Navy Yard. Other submarines, in various stages of completion, can be seen in the background. The O-class sub had a hull design that would have made Jules Verne proud. Amazingly, the vessel stayed in service training submarine crews until the end of World War II. (Puget Sound Navy Museum.)

On October 17, 1918, the Puget Sound Navy Yard launched submarine *H-7*. The vessel was originally intended to go to the Imperial Russian Navy, but after the Russian Revolution, it was acquired by the U.S. Navy, and it served the majority of its short career along the West Coast. *H-7* was decommissioned after only four years in naval service. (Puget Sound Navy Museum.)

At nearly every navy port, there were businessmen ready to cater to a sailor's needs—including providing a nice-looking, patriotic photograph postcard to send to that special someone. This photograph, taken by H. E. Yale in Bremerton, was sent to a young lady in Aurora, Oregon, in 1918. The sailor, posing against the stylized battleship and flag background, was an enlisted cook or baker aboard one of the vessels briefly visiting the Puget Sound Naval Yard. (Author's collection.)

At the time the above photograph was taken, on January 8, 1919, the vessel taking shape in the Puget Sound Navy Yard dry dock was simply known as "ammunition ship number one." By August 13, 1919, the ship was named USS *Pyro* (AE-1)—after the Greek word for fire. The explosives hauler departed Puget Sound in September 1920, but was decommissioned in Bremerton just four years later. However, as world tensions grew in 1939, the navy brought the *Pyro* back into service. After steaming all over the Pacific carrying tons of bombs, torpedoes, and bullets (as well as troops, supplies, and equipment), the ship was decommissioned again in Seattle on April 2, 1946. (Above, Puget Sound Navy Museum; below, National Archives/U.S. Navy.)

USS *Washington* (ACR-11) was launched in 1905 and served in the navy until 1916, when its name was changed to USS *Seattle* (CA-11). The name "Washington" would soon be assigned to a bigger and more modern battleship. The *Seattle* served during World War I as the flagship to a group of destroyers escorting convoys across the Atlantic. After the war, the ship periodically visited the city of its namesake during its cruises—always an exciting event for the citizens of Seattle. (Author's collection.)

At Puget Sound Navy Yard, Mike the cat is photographed atop a bollard on the bow of an unnamed navy ship. Sea cats have been mascots on vessels since ancient times. In practical terms, the feline hunters killed pests such as rats and mice. Some superstitious sailors believed that cats aboard their vessels always brought good luck. (Puget Sound Navy Museum.)

The first battleship named after Washington State never served the navy. The Colorado-class vessel was christened USS *Washington* (BB-47) and launched in 1921 in Camden, New Jersey. However, while the new ship was being fitted out in 1922, the United States and four other nations signed a treaty limiting naval arms and construction on the ship ceased. In 1924, the *Washington*, which was more than 75 percent complete, was towed out to sea and sunk as a target. (Puget Sound Navy Museum.)

As times change and technologies improve, naval vessels require constant maintenance and periodic major overhaul. For ships assigned to the Pacific, much of this work took place in Puget Sound Navy Yard. In this image, workers pose for a photograph after regunning the USS *Texas* (BB-35) in 1923. *Texas*, launched in 1912, served the navy through World War II, fighting in both the Atlantic and Pacific. (Puget Sound Navy Museum.)

USS *Tacoma* (then designated CL-20) was lost in a heavy storm near Vera Cruz, Mexico, in the Caribbean. The crew labored for nearly a week to free the *Tacoma* from Banquilla Reef. Four men, including the *Tacoma's* captain, drowned in the unsuccessful attempts. (Puget Sound Navy Museum.)

USS *Medusa* (AR-1) was the first purpose-built repair ship in the U.S. Navy. The vessel was constructed at Puget Sound Navy Yard from 1920 to 1924 and was outfitted with heavy equipment and large work spaces to upgrade and repair other warships in the fleet. More than 16 months of repair work fell in the *Medusa's* lap when the ship was moored in Pearl Harbor on December 7, 1941. It was April 1943 before the repair ship left Hawaii and moved out into the Pacific. After Japan's surrender, the well-worn *Medusa* was towed to Bremerton for decommissioning. Eventually, the veteran ship was scrapped in Portland, Oregon. (Puget Sound Navy Museum.)

On October 18, 1924, the USS *Shenandoah* (ZR-1) visited the army's Camp Lewis near Tacoma. On the clear fall day, thousands came out to see the navy's first rigid airship as it linked to a newly built 165-foot high mooring mast. The airship departed for San Diego the following day. The 680-foot long *Shenandoah* was lost less than a year later in a storm near Caldwell, Ohio. The Camp Lewis land where the mooring mast once stood is now part of McChord Air Force Base. (Fort Lewis Military Museum.)

USS *Gannet* (AM-41) was a minesweeper-turned-seaplane tender built at the end of World War I. The ship was often stationed in Alaska but steamed south to Sand Point in June 1926 to pick up a Loening amphibious aircraft. The *Gannet* was sunk by a German U-boat off Bermuda in June 1942. (National Archives/U.S. Navy.)

The unique-looking submarine tender USS *Holland* (AS-3) was named after John P. Holland, the inventor of the navy's first submarine. Built in Puget Sound Navy Yard in 1926, the *Holland* is seen here in the midst of its fitting-out process, on April 28, 1926. The sub tender was commissioned on June 1 of that same year. (Puget Sound Navy Museum.)

Three

The Golden Age

On July 4, 1928, the shipbuilding crews at Puget Sound Navy Yard begin work on the keel of "Cruiser 28." From this heavy nautical backbone, thousands of men would create the USS *Louisville*, which would be designated CA (heavy cruiser) No. 28. The honorary rigging gang is composed of some of the shipyard's most senior workmen. (Puget Sound Navy Museum.)

The battleship that would become famous in the opening minutes of America's involvement in World War II was a regular visitor to Puget Sound in the decades before Pearl Harbor. The first image shows USS *Arizona* (BB-39) during an upgrade at Puget Sound Navy Yard on February 3, 1928. The picture below shows the battleship's liberty boat, filled with Canadian sailors, during Fleet Week in Elliott Bay in 1934. In early 1941, the *Arizona* visited Puget Sound for the last time, receiving an overhaul in Bremerton before steaming back to Pearl Harbor. (Both, Puget Sound Navy Museum.)

America's second and third aircraft carriers began life as a pair of battle cruisers, but after the Washington Naval Conference, which limited the number and size of capital ships, the vessels were finished to serve a different purpose. The relatively new ships were photographed at Puget Sound Navy Yard in September 1928. USS *Lexington* (CV-2), moored on the left, and USS *Saratoga* (CV-3) were nearly identical except for a walkway across the funnel on the latter. (National Archives/U.S. Navy.)

The naval airfield at Sand Point was carved from a wooded foreland on the shores of Lake Washington. Mud Lake, seen above the gravel dredge in this 1928 photograph, was filled in as the navy air station expanded. Rumor has it that several pieces of earthmoving equipment were lost in the muddy bog before it was completely conquered. (National Archives/U.S. Navy.)

When the city of Tacoma ran out of power in the winter of 1929, civic leaders "borrowed" the power-generating abilities of the USS *Lexington* (CV-2). A drought pushed local hydroelectric power sources to their limits, leading to the unusual addition of "Lady Lex's" boilers to the city power grid. The electrical lifelines were photographed extending from the carrier's hull at Baker Dock on December 18, 1929, by a government photographer. The ship supplied Tacoma with one-quarter of its electrical needs for around one month until rain filled local reservoirs. (Above, Puget Sound Navy Museum; below, National Archives/U.S. Navy.)

This Boeing fighter was one of the most successful airplanes of the era, with over 580 of the type built in Seattle. The versatile airplane was made for the navy and designated the F4B, but the army adopted the fighter as well, calling it the P-12. This F4B-1, heavily loaded with bombs, was photographed at the company's hangar at Boeing Field on June 17, 1929. (National Archives/U.S. Navy.)

USS *Langley* (CV-1) was the first aircraft carrier in the navy. On the flight deck hastily built over the hull of the converted collier, some of America's first navy pilots learned to fly. This image was taken while the ship was in dry dock in Bremerton on January 7, 1930. (National Archives/U.S. Navy.)

During a visit to Bremerton in September 1931, the USS *Saratoga* (CV-3) was photographed rigged with ropes for painting its hull. Sailors often said that caring for the exterior of a navy vessel was a full-time job. The 880-foot-long aircraft carrier must have been a painter's nightmare. (National Archives/U.S. Navy.)

A quick photographer shot this image of a roving pack of feral dogs cruising the perimeter of one of the dry docks at Puget Sound Navy Yard in April 1932. Some jokester has added his own cartoon-like commentary to the scene directly onto the negative. (Puget Sound Navy Museum.)

The Bremerton-built USS *Holland* (AS-3) is seen here moored with a gaggle of S-class submarines in San Diego in the decade before World War II. When war came, the *Holland* served in the Pacific, repairing and refitting submarines relatively close to their patrol zones. (National Archives/U.S. Navy.)

In the 1930s, Puget Sound Navy Yard expanded greatly, thanks to works projects funded by Congress. The hammerhead crane was a distinctive part of the Bremerton skyline and was used for installing heavy guns in battleships. Though referred to as the "250-ton crane," the apparatus lifted more than 350 tons during tests, including four 14-inch guns, four 6-inch guns, and two hefty pieces of armor plate. (Author's collection.)

The oldest U.S. Navy warship to ever enter Puget Sound was the USS *Constitution*, launched in 1797. The newly restored "Old Ironsides" came to the Pacific Northwest during a three-year tour of 90 port cities in the United States. The famous frigate's stops in Puget Sound included Bellingham, Anacortes, Everett, Port Townsend, Bremerton, Seattle, Tacoma, and Olympia. This image was taken as a tug brought the veteran vessel into Port Angeles Bay on May 27, 1933. (Author's collection.)

Before workmen repainted or repaired a ship, they had to clean. Rust, worn paint, and barnacles covered the hulls of navy vessels with aggravating regularity. Here USS *Omaha* (CL-4) undergoes grooming at Puget Sound Navy Yard in 1933. A clean hull increased speed and lowered fuel consumption. (Puget Sound Navy Museum.)

The 600-foot-long USS *Louisville* (CA-28) was the largest ship built at Puget Sound Navy Yard. The vessel was equipped with nine 8-inch main guns and displaced over 9,000 tons. The *Louisville* steamed from Bremerton to New York City on its shakedown cruise in 1931. In wartime, the vessel became an important part of the navy's Pacific fleet. (Puget Sound Navy Museum.)

One of the ships of the Great White Fleet returned to Puget Sound in a different capacity in the years between the world wars. The navy decommissioned the battleship USS *Kearsarge* (BB-5) in 1920. At a shipyard on the East Coast, the ship's turrets and much of its superstructure were stripped away, clearing room for a massive revolving crane mast able to lift up to 250 tons. For added stability, workers created wide, 10-foot blisters affixed to either side of the vessel's hull. As "Crane Ship No. 1," the converted dreadnought transited the Panama Canal and worked for years in Bremerton, assisting with construction and repair, as well as regunning naval warships. After decades of service at naval yards on both coasts, the unique vessel was sold for scrap in 1955. (National Archives/U.S. Navy.)

Some of the earliest navy reports on Puget Sound describe the potential for storing ships in the nearby freshwater lakes. After World War I, Lake Union became a winter port for sailing ships and a floating graveyard for mothballed freighters. Poking fun at America's wartime president, locals called the rotting vessels "Wilson's Wood Row." The ships were a part of the Seattle scenery throughout the 1920s and into the 1930s. This image was taken after the 1932 opening of the George Washington Memorial Bridge (more commonly known as the Aurora Bridge). (Author's collection.)

Today the Boeing Company is known for building large multiengine aircraft. But in the years before World War II, the Boeing Airplane Company was one of the country's premier fighter plane builders. Here a three-ship flight of Boeing F2Bs, flown by the navy's High Hats performance squadron, cruises over the Florida coast near Pensacola in March 1933. Forerunners to the Blue Angels, the High Hats were known for taking off, looping, and landing with their planes tied together with short lengths of manila line. (National Archives/U.S. Navy.)

Flat plates and straight lines made an "Eagle Boat" quick and easy to build. Mass-produced by the Ford Motor Company near Detroit, Michigan, the steel vessels were the answer to the German U-boat threat during World War I. However, the first ship in the class was not launched until after the Armistice. *PE-57*, seen here in Bellingham Bay, was the third from the last Eagle Boat—launched on July 27, 1919. Unlike most of the ship's sisters, *PE-57* stayed in naval service throughout World War II, towing target spars for Whidbey-based torpedo bombers. (Jim Caldwell.)

A flight of navy observation aircraft cruises over Smith Tower in 1936. The 42-story building was the tallest structure west of the Mississippi River until 1931. Throughout the 1930s, U.S. Navy planes were most often painted white with bright, bold insignia and colors, making them stand out over the gray Seattle skyline. (Museum of History and Industry.)

When the U.S. Navy came into town, everyone knew it. This image was taken from West Seattle, looking east across Elliott Bay at the naval vessels in town for Fleet Week in July 1937. Similar displays took place throughout the 1930s near Seattle and Tacoma. The large spotlights made for a dazzling show, but their main function was to illuminate enemy aircraft and ships during night combat. (Author's collection.)

On an October day in 1938, Catalina patrol planes assigned to Naval Air Station Seattle (Sand Point) cruise over Lake Washington in celebration of Navy Day. The long-lived amphibious aircraft were a fairly common site in Seattle from the late 1930s to the 1950s. In wartime, the planes were often used to escort ships, fly long search missions, or snoop out enemy submarines. (National Archives/U.S. Navy.)

In the years before World War II, the naval air station at Sand Point grew considerably. This image from February 17, 1939, shows the construction of new seaplane hangars at the north edge of the base. Note the paved ramp at the water's edge, allowing amphibious aircraft like the Consolidated PBY Catalina to taxi from Lake Washington onto land. (National Archives/U.S. Navy.)

Still equipped with a distinctive 125-foot mooring mast left over from its first stint with the navy, USS *Patoka* (AV-6) was recommissioned as a seaplane tender in Bremerton in late 1939. The former oiler was used as an experimental base for dirigibles during the late 1920s and early 1930s before being mothballed. Outlasting the airships, it was modified to serve for more than a decade, the vessel worked in the Atlantic and Pacific throughout World War II, before finally being sold for scrap in 1948. This image was taken at Puget Sound Navy Yard on January 16, 1940. (National Archives/U.S. Navy.)

The second battleship named after the State of Washington was completed in May 1941 and spent the majority of the war in heavy combat operations in the Pacific. Amazingly, *Washington* (BB-56) was never hit by enemy fire. The battleship was, however, heavily damaged in a collision with the USS *Indiana* (BB-58) in early 1944 and sailed to Puget Sound Navy Yard to receive a new bow. (Puget Sound Navy Museum.)

When the cruiser USS *Omaha* (CL-4) was launched on December 14, 1920, it was the longest vessel to be built in the Pacific Northwest. It was also the first ship built in Puget Sound to slide down its slipway bow first when it was launched from the Todd Dry Dock and Shipbuilding Company's Tacoma shipyards. On November 6, 1941, the *Omaha* assisted in the capture of the disguised German freighter *Odenwald* in the Atlantic. After service in World War II, the *Omaha* was scrapped. (Author's collection.)

Four

World War II

Over the launching ways at Puget Sound Navy Yard, Mrs. A. M. Charlton awaits the word to smash a bottle of champagne over the bow of the new seaplane tender USS *Biscayne* (AVP-11). Aviation support ships of various types were built at PSNY in Bremerton, Seattle-Tacoma Shipbuilding Corporation in Tacoma, and at Lake Washington Shipyard in Houghton, Washington, during World War II. *Biscayne* was launched on July 3, 1941, and later became an amphibious force flagship, serving in the Atlantic, Mediterranean, and Pacific. (Puget Sound Navy Museum.)

In the early months of war, it seemed plausible that Japanese bombers might attack Puget Sound facilities, including sea and air bases, factories, and shipyards. Smoke generators helped obscure these potential targets from the air. The first image shows a smoke test at Puget Sound Navy Yard, conducted on March 10, 1942. A photography plane from NAS Seattle recorded the results from 5,000 feet for later evaluation. Attempts made months later, like the drill shown below on July 2, 1942, employed additional smoke generators on barges and boats to better obscure the yard's distinctive piers and dry docks. (Above, National Archives/U.S. Navy; below, Puget Sound Navy Museum.)

Fearing attack from the Japanese, the officials at Puget Sound Navy Yard ordered the construction of a number of bomb shelters near the piers and dry docks to protect workers during an air raid. The concrete-reinforced bunkers were constructed in mid-1942. This photograph, taken in the summer of 1943, shows one of the shelter entrances, thankfully hardly ever used. (National Archives/U.S. Navy.)

The cruiser USS *New Orleans* (CA-32) lost its bow to a Japanese torpedo during the Battle of Tassafaronga on November 30, 1942. With nearly one-quarter of the cruiser torn free (including the ship's foremost 9-inch turret), the *New Orleans* slowly steamed to the Solomon Islands for repairs and then to Australia for a makeshift "stub bow." Plodding slowly across the Pacific, the heavily damaged vessel arrived in Bremerton for permanent repairs in March 1943. A welded steel bow was added to the riveted steel ship and the *New Orleans* was back in the Pacific in the summer of 1943. (National Archives/U.S. Navy.)

Maybelle Wagner, welder at Puget Sound Navy Yard, poses for a photograph near a mass of steel that will become the keel of the destroyer escort USS *Wyman* (DE-38). The "38" seen in the image refers to the navy's numeric designation for the future ship. At first the ship was slated to be turned over to the British, but it was transferred back to U.S. Navy service before its completion. This image was taken on the ship's first official day in existence, September 7, 1942. (Puget Sound Navy Museum.)

Launched as USS *Kickapoo* at the end of World War I, the fleet tug USS *Mahopac* (AT-29) spent most of its life plying the waters near where it was built—Puget Sound Navy Yard. This photograph was taken on January 13, 1942. The tug served throughout World War II, pushing big battleships and carriers into the piers at the naval yard. (Puget Sound Navy Museum.)

USS *Bogue* (CVE-9) cruises into Commencement Bay in late 1942. The lead ship in its class of escort carriers, the *Bogue* was built at Seattle-Tacoma Shipbuilding Corporation in Tacoma, less than a mile away from where this photograph was taken. Airplanes from the ship assisted in the destruction of 13 German and Japanese submarines in combat operations during World War II. (National Archives/U.S. Navy.)

In typical pilot fashion, a flight instructor at Sand Point talks to a cadet pilot with his hands. The student wears a Gosport tube helmet, which allowed the instructor to critique and direct his student while the Ryan trainer plane was in flight. This photograph was taken in the summer of 1942. (National Archives/U.S. Navy.)

At the University of Washington, civilians take classes before joining the Volunteer Port Security Force (VPSF) on Seattle's piers and wharves. Protecting and patrolling the waterfront was a thankless job, but it was critical to the war effort. Sabotage, theft, or fire could harm the nation's ability to fight in the Pacific. Coast Guard managers of the VPSF program noted that most of the volunteers they recruited for the security jobs were slightly too old to be drafted, but still had a desire to serve in some capacity. (National Archives/U.S. Navy.)

During the first part of World War II, Seattle's waterfront was closed to civilian traffic. In order to keep the port open and safe for military activities, visitors had to show an identification card. Here two Coast Guard policemen examine a worker's pass at one of many waterfront checkpoints. (National Archives/U.S. Navy.)

Early in the war, it seemed completely possible that Japanese soldiers might slog ashore at sites on the West Coast of the United States. Puget Sound, along with California ports, was a likely location. With that in mind, Coast Guardsmen took their bayonet training drill quite seriously—their lives might depend on it. This image was taken at the Coast Guard's district training station at Friday Harbor on San Juan Island. (National Archives/U.S. Navy.)

It was always important to recruit more women into the Coast Guard's SPAR (women's reserve) program. The military and civilians worked together to provide engaging experiences for would-be new members. This image was taken during the first recruiting cruise, in June 1943. Prospective candidates were taken on a cruise of Puget Sound and treated to a picnic lunch on a scenic nearby island. (National Archives/U.S. Navy.)

The first members of the Coast Guard women's reserve to arrive in Puget Sound were trained in Cedar Falls, Iowa. Dubbed "SPARS," from the Coast Guard motto "Semper Paratus" and its English translation, "Always Ready," the women took on stateside roles that allowed more young men to fight overseas. This image of the "Original 19" was taken in Seattle at the end of February 1943. (National Archives/U.S. Navy.)

In continuing training, SPARs participate in a gas mask drill at Coast Guard Air Station Port Angeles in 1943. Many of the young women had brothers, fiancés, or husbands in the service and volunteered to do their part to hasten the war effort. (National Archives/U.S. Navy.)

Many of the jobs Coast Guard SPARs took over were clerical in nature. However, others were involved in nearly every non-combatant aspect of the military, including aviation. Here a pair of SPARs tows a Grumman Widgeon at Air Station Port Angeles. The amphibious planes were used in search and rescue efforts and their pilots helped patrol the Strait of Juan de Fuca in search of enemy submarines. (National Archives/U.S. Navy.)

In the Strait of Juan de Fuca, a pair of Grumman TBF Avenger bombers from NAS Whidbey Island makes a practice attack run on a destroyer in June 1943. Early in the war, the slow speed and low altitude required to properly launch a Mark 13 aerial torpedo exposed the attacking plane to murderous fire during combat conditions. (National Archives/U.S. Navy.)

Navy pilots viewed ferries like M/V (motor vessel) *Enetai* as targets. While the USS *Enterprise* (CV-6) was overhauled in Bremerton, flyers from the carrier calibrated their new search radars at night on the blacked-out ships cruising Puget Sound, making mock attacks just feet above the waves. Using the techniques they developed in Puget Sound, *Enterprise* pilots sank eight ships and damaged six more at Truk lagoon on February 17, 1944. (Steve Pickens.)

A Lockheed PV-1 Ventura from NAS Whidbey Island lets loose a practice torpedo over Puget Sound in June 1943. Pilots say that the big patrol bomber flew like a fighter—fast and maneuverable. The falling torpedo has a plywood box around its tail to slow and stabilize the weapon during its fall. Once it hit the water, the wooden shroud broke away. (National Archives/U.S. Navy.)

This image shows what were considered the "old" torpedo storage racks at NAS Whidbey Island in December 1943. With the war under way and the air station growing, there were new torpedo racks too. They were located outdoors and had room for up to 150 of the weapons, each weighing over a ton. (National Archives/U.S. Navy.)

To minimize the chance of setting off a magnetic mine, the navy acquired coastal minesweepers like the USS *Crow* (AMc-20), an ex-trawler with a wooden hull. But while being used to tow targets and recover practice torpedoes in Puget Sound, the *Crow* was vulnerable. In August 1943, an errant weapon, dropped from a navy bomber during exercises, struck the *Crow's* soft underbelly. The torpedo was inert but was running fast enough to punch a large hole in the 104-ton vessel, sending it to the bottom. (National Archives/U.S. Navy.)

Heading to Alaska on its sixth war patrol, *S-28* leaves Puget Sound after modifications in the summer of 1943. The old S-class submarine was laid down in 1919 and was still in service during World War II, patrolling the Northern Pacific. A year later, the submarine was lost during a training accident off Hawaii with the loss of its entire crew—50 men. (Puget Sound Navy Museum.)

Many of the battleships damaged in the attack on Pearl Harbor were repaired and refitted at Puget Sound Navy Yard, including USS *Maryland*, *Tennessee*, and *West Virginia*. One of the most heavily damaged vessels to be revived was USS *California* (BB-44), which arrived in Puget Sound in the summer of 1942. In these photographs, workers cut away bomb-damaged parts of the ship's superstructure. The *California* departed Bremerton in January 1944. (Both, National Archives/U.S. Navy.)

The Seattle-Tacoma Shipbuilding Corporation in the Tacoma tide flats became one of America's premier builders of escort carriers during World War II. Often dubbed "jeep carriers," the relatively small craft could be built cheaper and faster than larger types. Taken on April 6, 1943, this aerial image shows all eight slipways filled with carriers in various stages of construction and others moored nearby. (National Archives/U.S. Navy.)

USS *Sunset* (CVE-48) was launched into Commencement Bay on July 15, 1943. Built in Tacoma, the escort carrier was transferred to the UK that same year and became HMS *Thane* (D48). The ship survived an attack by a German U-boat in the Irish Sea in early 1945. Though towed to Scotland, the ship was declared a "constructive total loss" and was scrapped months later. (National Archives/U.S. Navy.)

USS *Tennessee* (BB-43) was damaged at Pearl Harbor and twice sailed to Puget Sound Navy Yard for improvements and repairs. This image shows the fully modified battleship near Port Angeles on May 12, 1943, after its second modernization. The ship's new torpedo bulges widened the battlewagon by 6 feet, making it too big for the Panama Canal. The *Tennessee* fought in the Pacific until 1945. (National Archives/U.S. Navy.)

Lake Washington Shipyard in Houghton, Washington, became a builder of AVP-class small seaplane tenders during World War II. Houghton is located south of Kirkland on the east shore of the lake. This photograph of the yard, taken in April 1943, shows many of the 311-foot vessels in various stages of construction. (National Archives/U.S. Navy.)

The Boeing Airplane Company worked to create an amphibious patrol bomber in the months before U.S. entry into World War II. Boeing's plant in Renton was slated for construction of the huge planes, designated PBBs and nicknamed Sea Rangers. But the navy's interest in the monster flying boat waned, and the Renton plant was turned over to the army for construction of B-29 Superfortress bombers. The prototype Sea Ranger, seen here in February 1943, went through its navy trials at Sand Point but was never put into production. The plane was the only one of its type to ever be built, and locals called it the "Lone Ranger." (National Archives/U.S. Navy.)

In September 1943, a pair of new Grumman F6F-3 Hellcat fighters cruise the skies over Puget Sound after takeoff from Sand Point. That same month, the first examples of the type debuted in combat in the Pacific. By the end of the war, Hellcat pilots had scored over 5,000 air-to-air victories over their Japanese foes, including the once-feared Mitsubishi Zero. (National Archives/U.S. Navy.)

At the south end of Lake Union lies the U.S. Naval Reserve Armory. Built from 1940 to 1942 as a Works Progress Administration project, the art deco building featured a large drill hall, which was used to train thousands of young U.S. Navy, U.S. Marine, and U.S. Coast Guard reservists during World War II. This aerial image was taken by a navy aircraft on April 28, 1943. (National Archives/U.S. Navy.)

A photographer from the USS *Long Island* (CVE-1) shot this image of a pair of the carrier's Curtiss SO3C Seamew aircraft during a visit to an army air base in Kitsap, Washington. Versions of the scout and observation plane were equipped with Ranger V-12 engines and built for operations off runways and carrier decks. Sailors considered them some of the homeliest aircraft to ever serve in the U.S. Navy. (National Archives/U.S. Navy.)

USS *Wyman* (DE-38) slides into Sinclair Inlet after launch on June 3, 1943. While originally destined to be transferred to the UK, the ship was retained by the U.S. Navy. In the Pacific, the *Wyman* was credited with sinking two Japanese submarines during escort duties. (Puget Sound Navy Museum.)

In the summer of 1943, destroyer escorts take shape in dry dock three at Puget Sound Navy Yard. The ships are USS *Brackett* (DE-41), USS *Reynolds* (DE-42), USS *Mitchell* (DE-43), and USS *Donaldson* (DE-44). All four served in the Pacific, and though one collided with a whale and another was damaged in a typhoon, the ships all survived the war. Each was retired from navy service soon afterwards. (Puget Sound Navy Museum.)

Off Point No Point near Whidbey Island, the USS *Donaldson* (DE-44) was photographed just 17 days after its commissioning at Puget Sound Navy Yard. In the Pacific, the diminutive *Donaldson* survived a typhoon that sank three larger destroyers. The destroyer escort later served in support of the invasions of Iwo Jima and Okinawa as well as in some of the final attacks on the Japanese mainland. (Puget Sound Navy Museum.)

The Coast Guard confiscated some of Puget Sound's privately owned craft during the war years. The *Electra* was designed by famous naval architect L. E. "Ted" Geary for the president of Puget Sound Power and Light. The beautiful 96-foot yacht was launched in 1930 at Lake Union Drydock. In wartime, the acquired vessel was slathered in gray paint and pressed into service as a patrol craft. (National Archives/U.S. Navy.)

At Neah Bay (the northwest tip of Washington State) a pair of Coast Guard craft stands ready for action. The vessel in the foreground is a 38-foot picket boat, made for patrolling and policing harbors. It was built in Kirkland, Washington, at Ballinger Boat Works. The craft in the background is a 63-foot air-sea rescue boat, commonly dubbed a "crash boat." It could quickly assist in emergencies, including rushing out to retrieve downed airmen off the coast. (National Archives/U.S. Navy.)

There was hardly ever a traffic jam in Seattle until World War II. But this is the scene at the corner of Duwamish Avenue and Spokane Street as wartime ship workers move onto Harbor Island during a shift change. With the Boeing Airplane Company churning out bombers at Boeing Field and Seattle-Tacoma Shipbuilding Corporation building destroyers a few miles away, the south end of Seattle was a very busy place. (National Archives/U.S. Navy.)

During a two-month overhaul in the fall of 1944, workmen at Puget Sound Navy Yard cut into the deck of USS *Yorktown* (CV-10) in order to install an improved aircraft catapult. By November 1944, the carrier was back in combat, launching air strikes against targets in the Philippines. (National Archives/U.S. Navy.)

It was most likely a careless antiaircraft gun crew that started a fire at Seattle's Pier D on April 26, 1944. The soldiers' electrical cooking appliances or oil heaters were blamed for the blaze that engulfed the warehouse filled with war commodities. In this photograph, Coast Guard fire barges as well as Seattle's fireboats *Alki* and *Duwamish* (the latter of which was manned by the Coast Guard during World War II) frantically pump water onto the waterfront fire. (National Archives/U.S. Navy.)

USS *Ross* (DD-563) was one of 21 Fletcher-class destroyers built at Seattle-Tacoma Shipbuilding Corporation at Harbor Island. The ship joined the naval forces attacking the Marianas Islands in June 1944 and was damaged by two mines off the Philippines in October of that same year. After years of reserve duty, the *Ross* was sunk as a target off Puerto Rico in 1978. (National Archives/U.S. Navy.)

The Tacoma-class patrol frigate USS *Pocatello* (PF-9) was manned by a Coast Guard crew and spent much of World War II patrolling in the Pacific about 1,500 miles west of Seattle. In December 1944, a crewman aboard the *Pocatello* suffered a ruptured appendix while the ship was 200 miles from Washington. Coast Guard flying boats attempted water landings to save the sick man, but the seas proved to be too rough. Two airdrops of intravenous fluids and penicillin via parachute stabilized the sailor and kept the *Pocatello* on station. (National Archives/U.S. Navy.)

During a practice drop over Puget Sound, this aerial torpedo failed to release from the belly of a Whidbey Island–based Lockheed patrol bomber. When the crew finally wrenched the errant weapon loose, it fell in a farmer's field on Whidbey. The flyers joked that they were the only aviators to ever torpedo a cornfield. Navy men later went out and retrieved the smashed weapon from the crater it left in the field. (National Archives/U.S. Navy.)

After pounding enemy positions on Guam for nine days in July 1944, the USS *New Mexico* (BB-40) spent August through October at Puget Sound Navy Yard for overhaul. Here the battleship receives new 14-inch guns for its forward turrets. (National Archives/U.S. Navy.)

In the summer of 1943, the USS *Abner Read* (DD-526) lost its stern to an explosion near Kiska Island in the Aleutians. Some 70 men were killed or missing in the blast, later attributed to a mine. After temporary repairs, the destroyer was towed to Puget Sound Navy Yard, where a new stern was grafted onto the ship. On November 1, 1944, *Abner Read* was sunk by a kamikaze attack off Samar Island in the Philippines. (National Archives/U.S. Navy.)

The ocean tug USS *Ute* (AT-76) served in Alaska and around the Pacific during World War II with a brief respite in Puget Sound for repairs and alterations in the summer of 1944. Here the *Ute* departs Bremerton, bound for Kodiak, Alaska, on July 19, 1944. The venerable *Ute* went on to serve in Korea, Vietnam, and then with the Coast Guard until retirement in 1988. Note the *Kalakala* docked at the ferry terminal in the background. (Puget Sound Navy Museum.)

With Puget Sound Navy Yard's massive hammerhead crane in the background, M/V *Kalakala* heads east towards Seattle, carrying another load of sailors and shipyard workers. Four ferries, including the sleek art deco flagship of the Black Ball fleet, handled the busy route, making 29 trips a day. From the Pearl Harbor attack until the Japanese surrender, Black Ball ferries safely carried 34,931,926 passengers and 5,795,761 vehicles across Puget Sound. (Author's collection.)

The Seattle-built USS *Jarvis* (DD-799) was named after James C. Jarvis, a 13-year-old midshipman killed in 1800 aboard the USS *Constellation*. The destroyer served with the U.S. Navy during World War II and Korea before being transferred to the Spanish navy in 1960. (National Archives/U.S. Navy.)

At Puget Sound Navy Yard, a party of VIPs poses for the cameras on May 4, 1944. The woman holding the bouquet at the center is Inez Cowdrey, the sponsor of the destroyer USS *Killen* (DD-593). During the Battle of Surigao Strait, the ship assisted in the sinking of the Japanese battleship *Yamashiro*. USS *Killen* was sunk as a missile target near Vieques Island, Puerto Rico, in 1975. (National Archives/U.S. Navy.)

Though labeled a Japanese suicide boat, the small vessel brought ashore at Puget Sound Navy Yard was a Daihatsu-style landing craft with a simple diesel engine, capable of carrying enemy troops, horses, and medium tanks. The captured 8-ton craft was of special interest to the boatbuilders working at the navy yard. Early versions were made of steel, but due to materials shortages, later boats were constructed of plywood and had oak frames. This image was taken on May 17, 1944. (Puget Sound Navy Museum.)

USS *West Virginia* (BB-48) was raised off the shallow bottom at Pearl Harbor and sailed to Puget Sound. At Bremerton, the battleship underwent an extensive rebuild and heavy modification. Seen here, the new "Wee Vee" departs Puget Sound in July 1944 on its way back to combat in the Pacific. The vessel wears a stunning dazzle paint scheme developed by the U.S. Navy Bureau of Ships. (National Archives/U.S. Navy.)

The commissioning program for the USS *Commencement Bay* (CVE-105) states, "The ship's name is unique in that she was built, launched, and commissioned on the bay from which she gets her name." When construction of the vessel started in Tacoma, it was called *St. Joseph Bay*, named after a body of water in Florida. But the citizens of Pierce County raised enough money in war bond drives to buy the escort carrier and rename it after the local waterway. (Puget Sound Navy Museum.)

A No. 800 Squadron Hellcat of the Royal Naval Air Service touches down on the deck of the HMS *Emperor* (D98) in September 1944. The Bogue-class escort carrier was built in Tacoma and later turned over to the UK as part of lend-lease agreements between Britain and the United States. During combat in Europe, planes from the ship participated in many missions, including D-Day operations and the invasion of Southern France. (Grumman History Center.)

Naval Auxiliary Air Station Shelton was an auxiliary airfield to NAS Seattle. In the lower part of this image is a large collection of General Motors FM-2 Wildcat fighters from composite squadrons readying for combat duty in the Pacific. The other side of the ramp shows the tow and utility planes of Utility Squadron 13. Each large two-engine plane is a Martin JM-1, a naval version of the B-26 Marauder bomber. These speedy planes towed target sleeves that were used to train both fighter pilots and antiaircraft gunners on the ground. (National Archives/U.S. Navy.)

At Seattle's Edison Vocational School, an art student puts the finishing touches on a new Coast Guard Volunteer Port Security Force recruiting poster. Around Puget Sound, scores of civilian schools, clubs, and organizations turned out to do their part for the war effort. (National Archives/U.S. Navy.)

The crew of the USS *St. George* (AV-16) fuels its first PBY patrol bomber in the sheltered waters of Puget Sound near Whidbey Island. The *St. George* was built in Tacoma and commissioned just days before this image was taken on August 9, 1944. (National Archives/U.S. Navy.)

Long before Bangor became a Trident submarine base, it was the site of a U.S. Naval Magazine. Beginning in early 1945, cargo ships would visit the isolated spot on Hood Canal and take aboard tons of ammunition. The bombs and shells were used to supply the vessels and aircraft operating in the Pacific through the last months of World War II and throughout the conflicts in Korea and Vietnam. This photograph was taken in November 1945. (National Archives/U.S. Navy.)

Seattle-Tacoma Shipbuilding Corporation, at the mouth of the Duwamish River, was the nation's third largest builder of destroyers during World War II. The 17,000 workers built and repaired an endless stream of ships during the war years at Harbor Island, just southwest of downtown Seattle. (National Archives/U.S. Navy.)

At a dry dock on Harbor Island, Seattle-Tacoma Shipbuilding Corporation workers overhaul a long-neglected liberty ship. As some men work to scour the barnacle-encrusted hull, others prepare to overhaul the cargo ship's massive 18-foot propeller. (National Archives/U.S. Navy.)

These images show the extent of the damage to the USS *Saratoga* (CV-3) after it was attacked by kamikaze pilots near Iwo Jima in February 1945. The above image shows a huge hole in the flight deck where an enemy aircraft crashed amid fully armed and fueled night fighters preparing to take off. The second image shows incinerated Hellcat fighters parked on the *Saratoga*'s hangar deck. The workers at Puget Sound Navy Yard stated that the carrier suffered the most extensive damage they had ever been tasked to repair. (Both, National Archives/U.S. Navy.)

Slathered in gallons of gray paint, USS *Saratoga* (CV-3) cruises through Puget Sound under dreary skies on May 15, 1945. At the time, the *Sara* was the nation's oldest surviving aircraft carrier and would soon become the only fleet carrier to serve in World War II from beginning to end. After the war, the old flattop was intentionally sunk by an atomic bomb blast at Bikini Atoll. (National Archives/U.S. Navy.)

Amid heavy bursts of antiaircraft fire, a Japanese bomber smashes into the Bremerton-built heavy cruiser USS *Louisville* (CA-28) in January 1945. The image was taken from a nearby U.S. aircraft carrier. That day in the Lingayen Gulf, two kamikazes hit the "Lady Lou." They would not be the last. After repairs, the *Louisville* was struck by another Japanese suicide plane off Okinawa in June 1945. (National Archives/U.S. Navy.)

On April 15, 1945, the USS *Laffey* (DD-724) was attacked by swarms of kamikaze aircraft near Okinawa. The destroyer miraculously survived repeated attacks from the air but was terribly damaged. After receiving quick repairs in the Pacific, *Laffey* was sent to Puget Sound for much more extensive work. In Seattle and Tacoma, "The Ship That Would Not Die" was toured by thousands of curious civilians. These images show the *Laffey* temporarily moored at Seattle's Pier 48 in the summer of 1945. (Both, Patriots Point Naval and Maritime Museum.)

Aboard the *Laffey* in Seattle, civilians got a chance to see for themselves what the fierce fighting in the Pacific was like for sailors. The navy used the damaged destroyer to remind the people back home that the war was considered far from over. The *Laffey* was "riddled like a sieve above the water line"—struck by four bombs and six kamikaze planes, it was strafed from stem to stern. Some 32 navy men were killed in the attacks and 71 more were wounded. (Patriots Point Naval and Maritime Museum.)

Though named after the home city of Puget Sound Navy Yard, USS *Bremerton* (CA-130) was built in New Jersey at the end of World War II. The cruiser completed two tours during the Korean War, using its 8-inch guns to support United Nations ground forces in 1951 and 1953. (Author's collection.)

A navy man in a staff car dutifully stops short of the active runway to let a Consolidated PB4Y-2 Privateer patrol bomber rumble in for a landing. NAS Whidbey's land plane airfield was named after William B. Ault, a flyer who went missing in action during the Battle of the Coral Sea. This photograph was taken on June 28, 1945. (National Archives/U.S. Navy.)

This Grumman F6F-5 Hellcat fighter, assigned to VF-35, ran into trouble on Whidbey Island on May 26, 1945. The Arlington-based pilot, who had never flown this type of aircraft before, was unpleasantly surprised by the amount of torque produced by the fighter's engine on takeoff. The plane's wing scraped along the runway and then the plane "ground looped," coming to rest upside down. The stunned aviator was able to wiggle from the cockpit, suffering only a scrape on his head. (Grumman History Center.)

This photograph of Naval Air Station Seattle was shot directly downward from a navy patrol plane cruising over Sand Point at 12,000 feet on April 4, 1945. The small crosswind runways and the shorter main runway were built first. Later a longer main runway was created parallel to the first. The latter crosses the base from side to side, taking full advantage of every inch available for the big four engine patrol bombers fielded during the war. (National Archives/U.S. Navy.)

At NAS Shelton, ground crews work to secure a Goodyear K-class blimp to its mooring mast in 1945. *K-71* was assigned to blimp squadron (Blimpron) 33, based in Tillamook, Oregon. The non-rigid airships were used to patrol the Pacific Coast, escorting ship convoys and searching for signs of enemy submarines. (National Archives/U.S. Navy.)

Workers at Puget Sound Navy Yard constantly updated and improved naval vessels serving in the Pacific during World War II. Near the end of the fighting, nearly every navy combat ship bristled with guns in an attempt to stop kamikaze suicide attacks. Here a worker rebuilds an endless stream of air-cooled 40-mm Bofors guns. (Puget Sound Navy Museum.)

Near Okinawa on May 10, 1945, a pair of Japanese kamikazes eluded patrolling fighters and a hail of antiaircraft fire to plunge into the flight deck of the aircraft carrier USS *Bunker Hill* (CV-17). Badly damaged, *Bunker Hill* limped eastward to Pearl Harbor and then on to Bremerton. This image was taken after the ship's arrival at Puget Sound Navy Yard on June 16, 1945. Ready to begin repairs, ship workers survey the mangled steel and shattered wooden planks of the *Bunker Hill*'s flight deck. The carrier was still under repair when World War II ended. (National Archives/U.S. Navy.)

Five

The Cold War

After fighting off the Marshalls, Marianas, Palaus, Leyte, and Mindoro, participating in battles in the Philippine Sea and Leyte Gulf, and launching air strikes over the Central and Western Pacific, the USS *Essex* (CV-9) finally returned to the United States, dropping anchor in Puget Sound on September 13, 1945. The battered bow reflects the carrier's hard life in World War II; commissioned on the last day of 1942, the *Essex* was in nearly constant combat operations. (National Archives/U.S. Navy.)

Puget Sound Navy Yard workers stream out of the half-repaired USS *Bunker Hill* (CV-17) on August 14, 1945. News had reached the docks that Japan had surrendered and the yard's foremen decided to let the overworked men and women leave the job early to celebrate the end of World War II. (National Archives/U.S. Navy.)

Too late for battle, the USS *Tinian* (CVE-123) was christened by Grace L. Woods just three days after the Japanese surrendered. The partly completed escort carrier slid down the launchways at Todd Pacific Shipyards in Tacoma on September 5, 1945. The ship was never commissioned and went directly into the navy's reserve fleet at Tacoma. The *Tinian* served as a helicopter ship, cargo vessel, and aircraft ferry before it was scrapped in 1971. (National Archives/U.S. Navy.)

A new 40-foot motor launch takes shape in Puget Sound Naval Shipyard's boat and joiner shop in October 1945. The builders were particularly proud of the laminated teak capping stringers and clamps. The boat could carry up to 75 men and weighed slightly over 8 tons. (Puget Sound Navy Museum.)

The destroyer tender USS *Isle Royale* (AD-29) was caught up in shipyard reshuffling at war's end. The above photograph shows the completed hull on launch day in Seattle on September 15, 1945. Soon after, the ship was transferred to Todd Pacific Shipyards in Tacoma for completion. Amazingly, it was 1962 before the ship was commissioned in the U.S. Navy. (Both, National Archives/U.S. Navy.)

Civilians tour the flight deck of the USS *Ticonderoga* (CV-14) on Navy Day. The aircraft in the foreground are, from left to right, a Curtiss Helldiver dive bomber, a Grumman Hellcat fighter, and an Avenger torpedo bomber designed by Grumman and built by General Motors. Both *Ticonderoga* and the battleship USS *Maryland* (BB-46), seen in the background, were on hand in Tacoma for the end-of-war Navy Day on October 27, 1945. (National Archives/U.S. Navy.)

A young girl takes her turn looking down the sights of the twin .30-caliber machine guns mounted in the rear cockpit of a Curtiss SB2C Helldiver dive bomber on Navy Day. The airplane was one of many naval craft on the flight deck of the USS *Ticonderoga* (CV-14) when the ship moored in Tacoma in October 1945. (National Archives/U.S. Navy.)

On the day after Christmas in 1945, Boeing test pilot Bob Lamson was flying an experimental Boeing naval fighter, designated XF8B-1. Swooping down on a pair of Sand Point Corsairs near Everett, Lamson was going to "pour on the coals" and show off. But when he pushed the throttle to full power, the fighter's monstrous R-4360 engine died. Lamson brought the stricken plane down at Everett Municipal Airport, where it nosed in on the soft ground. (Glenn Humann.)

The dry docks and piers at Puget Sound Naval Shipyard are packed with fighting ships in this aerial photograph taken from a passing navy aircraft on December 14, 1945. One can easily pick out the largest vessels—the Essex-class aircraft carriers *Bunker Hill* and *Essex*, as well as the Iowa-class battleships *New Jersey* and *Iowa*. (National Archives/U.S. Navy.)

The attack transport USS *Latimer* (APA-152) departs Seattle, passing by the U.S. Navy Supply Depot at Smith Cove in Elliott Bay in late October 1945. Seen moored at Pier 91 is USS *Iowa* (BB-61) on the right and the "old battlewagon," USS *Colorado* (BB-45), on the left. The battleships were launched more than 20 years apart—the *Colorado* in 1921 and the *Iowa* in 1942. (National Archives/U.S. Navy.)

Scores of navy vehicles and equipment were no longer needed by the postwar navy after the fighting ceased and the frantic pace of building wound down. This image taken on January 10, 1946, shows lines of dock mules and forklifts at the naval depot in Tacoma. Each machine is tagged and readied to be sold to the general public. (National Archives/U.S. Navy.)

"Boy is he happy," reads the caption that accompanies this image. Former Seabee (member of a navy construction battalion) Joseph Hickey took advantage of the influx of navy equipment flowing in from the Pacific battlefields after the fighting ceased by buying this 2.5-ton truck for $1,285. As he pulls out of the main gate of the South Tacoma Navy Materials Distribution Center, he pauses to shake the hand of the base's commanding officer. Hickey made his purchase in February 1946. (National Archives/U.S. Navy.)

Masses of military equipment flow back through Puget Sound's ports after the Japanese surrender. A typical scene in the summer of 1946 shows rows of Euclid construction tractors, laundry trailers, and even a Sherman tank. This image was probably taken at Pier 36 on the Seattle waterfront. (National Archives/U.S. Navy.)

Sometime after World War II, a flight of Vought F4U-1D Corsairs cruises near Seattle. The fighters, belonging to the Naval Air Reserve Training Unit based at Sand Point, were glossy sea blue with international orange numbers and stripes. A Sand Point Corsair that ditched in Lake Washington after a training accident was recovered in the 1980s and now resides, fully restored, at Seattle's Museum of Flight on Boeing Field. (National Archives/U.S. Navy.)

When crews emptied dry dock three at Puget Sound Naval Shipyard, they found much of the floor covered in nearly 3 inches of herring. After quickly gathering up a few of the fish for bait, the workmen refilled the basin and then used air hammers, lowered on ropes, to chase the fish out of the dock enclosure. The strange incident happened on January 18, 1949. (Puget Sound Navy Museum.)

An Avenger torpedo bomber ran into trouble while landing at Boeing Field in 1946. When the plane's Wright Cyclone engine began to sputter, the pilot, P. O. Hull, did his best to safely glide the plane down among the houses in the South Park neighborhood. The landing was not exactly picture perfect, wrecking a residence on Cloverdale Street. True to the adage that any landing one can walk away from is a good one, the pilot was unhurt in the accident. (National Archives/U.S. Navy.)

Naval aircraft from Sand Point fly overhead at the grand opening of Seattle-Tacoma International Airport's main terminal. The first commercial flights departed from Sea-Tac in 1947 and the new terminal opened to the flying public on July 9, 1949—dubbed "Conqueror's Day." The aircraft in this photograph are all World War II vintage: Vought Corsair fighters (right), Grumman Avenger torpedo bombers (center), and Grumman Hellcat fighters (lower left). (Museum of History and Industry.)

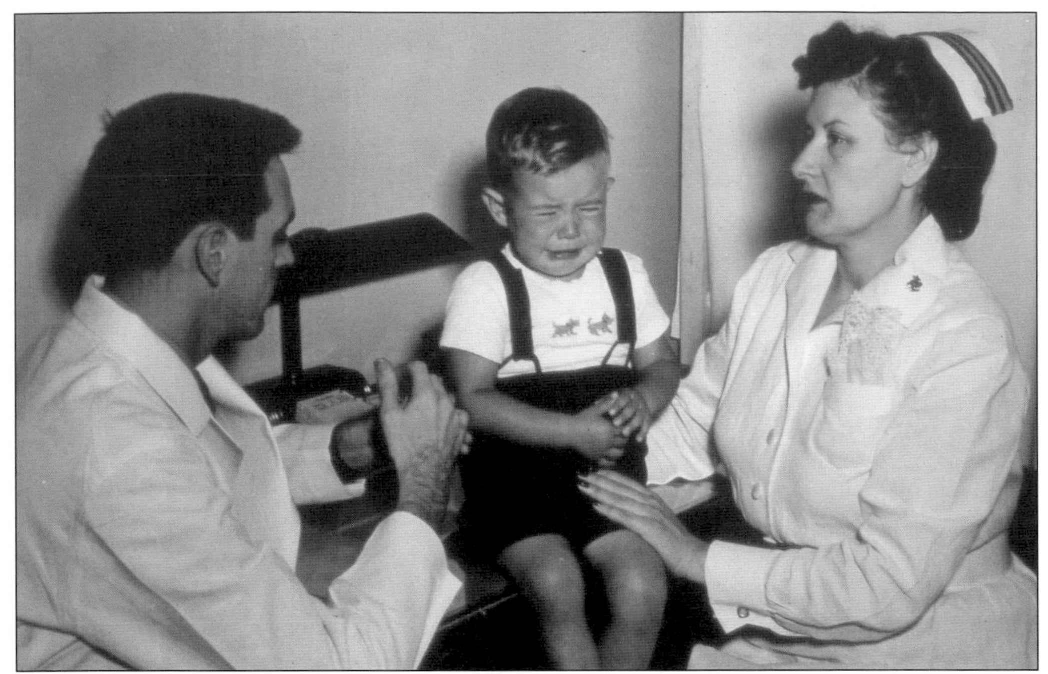

At a navy and marine family clinic in Seattle, the son of a navy commander reluctantly receives a shot from a doctor. This photograph was taken on October 4, 1950. (National Archives/U.S. Navy.)

Wives and girlfriends wave goodbye as the USS *Charles E. Brannon* (DE-446) departs Seattle. The destroyer escort operated as a reserve training ship after World War II. This image was taken in June 1950 when the ship was off to sea for training near Acapulco, Mexico. (National Archives/U.S. Navy.)

U.S. Marine Corps officers and men form up to march down the platform at Seattle's King Street train station. Their unit will soon be boarding a troop ship bound for the fighting in Korea. During both World War II and the Korean War, hundreds of thousands of military personnel from throughout the United States passed through the Puget Sound region on their way to combat zones in the Pacific. This image was taken on August 9, 1950. (National Archives/U.S. Navy.)

Tugs work to counteract the effects of the wind against the side of the USS *General* M. C. *Meigs* (AP-116) as soldiers board the troop transport at Seattle's Pier 90 on November 21, 1950. The U.S. Army soldiers were bound for fighting in Korea. The *Meigs*, long since retired, was lost while being towed near Cape Flattery at the mouth of Puget Sound in 1972. (National Archives/U.S. Navy.)

A valuable airplane and irreplaceable pilot cannot be risked to test a new catapult. To iron out the bugs, crewman repeatedly fired dead load trolleys from the flight deck of the USS *Essex* (CV-9) in February 1951. These tests took place just weeks after the ship's official recommissioning at Puget Sound Naval Shipyard. The newly modernized carrier was destined for combat operations off Korea. (Both, National Archives/U.S. Navy.)

A harbor tug wrangles a trio of heavy dead load trolleys from the waters in front of the bow of the USS *Essex* (CV-9) after catapult tests. With the carrier safely moored at Puget Sound Naval Shipyard, the trolleys could be quickly hoisted onto the flight deck for more tests. (National Archives/U.S. Navy.)

One of the most famous carriers of World War II, USS *Essex* (CV-9) was stored "in mothballs" at Puget Sound Naval Shipyard from early 1947 to 1950. In this image, taken on November 7, 1950, work continues to modernize the ship in order to enable it to launch and recover new jet-powered navy fighter and bomber aircraft. Later the *Essex* would earn four battle stars for its Korean War service. (National Archives/U.S. Navy.)

The World War II–era Essex-class carrier USS *Bon Homme Richard* (CV-31) was taken out of service at Puget Sound Naval Shipyard in early 1947. When the war in Korea flared up, the vessel was recommissioned in January 1951. These images, taken in February of that same year, show sailors preparing the veteran vessel for patrolling the Pacific yet again. After leaving Puget Sound, the carrier launched major attacks over Korea. (Both, National Archives/U.S. Navy.)

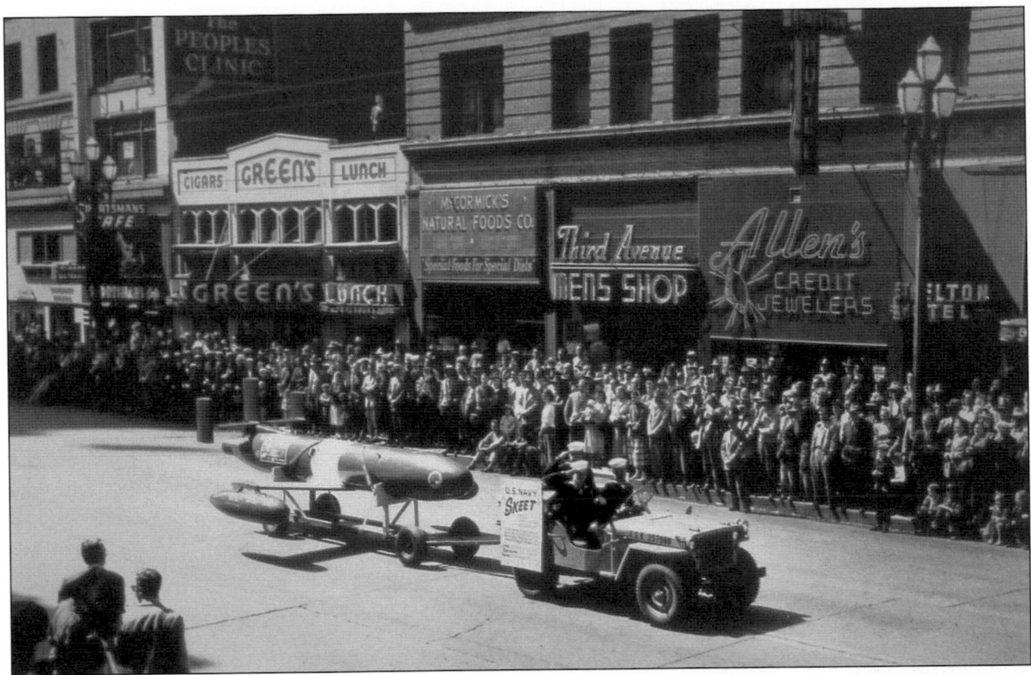

Sailors tow a rare Skeet missile down Seattle's Third Avenue during the Armed Forces Parade in 1951. The Curtiss KD2C-2, nicknamed Skeet, was an air-launched drone powered by a loud, fuel-guzzling, pulse-jet engine. The expendable unmanned craft were built to train warship gunnery crews in the art of knocking out attacking Soviet aircraft. Due to their low speeds and high rate of fuel consumption, few Skeets were produced before the program was cancelled in 1949. (National Archives/U.S. Navy.)

Aboard the destroyer escort USS *George* (DE-697), the Junior Seafair Royalty of 1951 raise their milk glasses in a toast at the beginning of Seafair—Seattle's summer festival. The kids were chosen by Seattle's Chamber of Commerce. They were dubbed king, queen, and junior pirate for the season. (National Archives/U.S. Navy.)

After the police discovered illegal slot machines near Seattle in early 1951, they staged a photo opportunity for local newspapers by eagerly smashing the banned gambling devices with a 5-pound sledge hammer. After the carnage, the navy assisted in the effort by loading the battered slots onto a tug and dumping them in Elliott Bay. (National Archives/U.S. Navy.)

In the summer of 1951, a PB4Y-2 bomber returns to NAS Seattle after a tour of duty patrolling the shores off Korea. After endless hours of flying back across the lonely Pacific, the crew of the tall-tailed Privateer happily wave at their welcoming committee from the upper turret and access hatches. The plane was assigned to VP (Patrol Squadron) 772. (National Archives/U.S. Navy.)

A large piece of the mangled remains of a Grumman F8F-1 Bearcat reveals the home base of this navy plane—Seattle. Flying from Sand Point, the speedy fighter crashed into the woods 5 miles southeast of Paine Field on February 8, 1951. The pilot was killed in the accident. The plane's big R-2800 engine, still encased in its battered blue cowling, can be seen on the right side of the frame. (National Archives/U.S. Navy.)

The crewmen of the USS *Valley Forge* (CV-45) celebrate their arrival in Elliott Bay with a special formation on the carrier's flight deck. Returning from combat operations in the waters off Korea, the carrier is being used to transport a number of navy aircraft, including two Martin PBM flying boats and a Consolidated PB4Y-1 patrol bomber. The *Valley Forge* came to Seattle in April 1951. (National Archives/U.S. Navy.)

Members of the National Editorial Association get a short cruise on the destroyer escort USS *Gilligan* (DE-508) in Sinclair Inlet near Bremerton on June 5, 1951. During World War II, a kamikaze crashed directly into one of the *Gilligan's* 40-mm gun tubs, killing 12 men and wounding 12 more. Amazingly, the vessel survived the massive fires that followed. Later the *Gilligan* was struck by an enemy torpedo—which did not explode. The veteran destroyer escort later became a reserve ship based in Puget Sound. (National Archives/U.S. Navy.)

In the summer of 1951, USS *Volador* (SS-490) transits the Ballard Locks on its way to a mock navy amphibious assault in Lake Washington. With a pyrotechnic charge, the submarine announced the beginning of the attack. The *Volador* served in the Pacific through the Korea and Vietnam eras before it was transferred to Italy in 1972. (National Archives/U.S. Navy.)

On August 3, 1952, a marine LVT-3C rumbles ashore at Sand Point during amphibious assault practice. The LVT (Landing Vehicle, Tracked) was similar to those used during the Battle of Inchon in Korea in 1950. (National Archives/U.S. Navy.)

The children of Duwamish Home Orphanage got quite a treat when the *North Pole Express* brought Santa (and a load of presents) to Naval Air Station Seattle on Christmas Eve Day, 1952. The *Express* is one of the navy's amphibious Consolidated PBY Catalina patrol bombers, specially painted snow white and adorned with cartoon and movie characters. (National Archives/U.S. Navy.)

Nearly every summer since 1952, the U.S. Navy's aerial demonstration team—the Blue Angels—has dazzled thousands of spectators over Seattle's Seafair celebration on Lake Washington. The core of the group flew combat missions in Korea before returning to stunt flying in 1951. Shown here in August 1954, some of the flyers pose near one of their dark blue and gold Grumman F9F-5 Panther jet fighters. (National Archives/U.S. Navy.)

A young sailor briefs John H. Turpin on the controls in the cockpit of a North American SNJ advanced trainer aircraft operating from Sand Point in 1952. At the time, "Dick" Turpin was the last survivor of the explosion and sinking of the USS *Maine* in 1898. On that fateful evening, the young mess attendant dove overboard into Havana Harbor. He later settled in Seattle, where he became a critical part of the region's naval community until his death in 1962. (National Archives/U.S. Navy.)

Young pilot Ens. James J. McGinnis is greeted by his mother and father at NAS Seattle after a long cross-country flight. The flyer and his Grumman F9F Panther fighter were most likely on their way to join a carrier air group headed to the Pacific to participate in combat operations over Korea. (National Archives/U.S. Navy.)

On a rainy August day in 1953, a Coast Guard PB-1G search plane skidded off the runway at NAS Seattle and splashed down in Lake Washington. The plane is a converted World War II–era B-17 Flying Fortress, probably built at Boeing's Plant 2 in south Seattle. After the war, some of these bombers were converted to lifeboat-carrying rescue planes. (National Archives/U.S. Navy.)

USS *Shangri-La* (CV-38) heads south in Puget Sound during sea trials after modernization in Bremerton in 1953 and 1954. The World War II veteran received new aircraft launch and recovery gear, including an angled deck for jet planes. The carrier's unusual name dates back to World War II, when the press asked President Roosevelt about the starting point for the aircraft that carried out the Doolittle Raid. Although the American bombers flew from the carrier *Hornet* (CV-8), Roosevelt told the reporters the planes came from the mythical world of Shangri-La. (National Archives/U.S. Navy.)

Horseplay and hazing were not the only character-building activities in the navy's 90-day accelerated summer program in Seattle, but they are what the photographers liked to shoot. In the first image, high school-aged recruits are tossed into Cottage Lake during a picnic to kick off the 1954 season. Weeks later at NAS Seattle, young men caught causing a disturbance in the ranks are ordered to "duck walk" around the formation of future sailors. (Both, National Archives/U.S. Navy.)

The ocean minesweeper USS *Guide* (AM-447) awaits send off at the Todd Pacific Shipyards in Seattle on April 17, 1954. By the time the wooden vessel was commissioned in 1955, its designation had been changed to MSO-447. *Guide* served its entire career in the Pacific, including combat patrols off the coast of Vietnam. The vessel was given up in 1974 and sold for $26,601 to a California scrap dealer. (National Archives/U.S. Navy.)

Shore patrolmen expel a pair of wild ducks from the front gate of NAS Seattle in April 1954. While most people—sailors and civilians—respect the imposing sentries at the entrance of the naval base, these wild waterfowl do not seem to care. (National Archives/U.S. Navy.)

The famous battleship USS *Missouri* (BB-63) was the site of the Japanese surrender in Tokyo Bay at the end of World War II. After fighting in the Korean War, the *Missouri* was stored at Bremerton beginning in 1955, becoming an iconic resident of Puget Sound Naval Shipyard. The ship was recommissioned in 1986 and later fought in Operation Desert Storm. Today it is moored at Pearl Harbor, Hawaii. (Puget Sound Navy Museum.)

After World War II, Korea, and Vietnam, Bremerton became the holding point for many warhorse Essex-class aircraft carriers. The above image shows the USS *Bon Homme Richard* (CV-31) along with other "mothballed" vessels at Puget Sound Naval Shipyard. The ship was scrapped in 1992. The second photograph captures a worker spraying preservative on the flight deck of the USS *Oriskany* (CV-34). The carrier was intentionally sunk off the coast of Florida in 2006 to create an artificial reef habitat for ocean wildlife. (Both, Puget Sound Navy Museum.)